I0824873

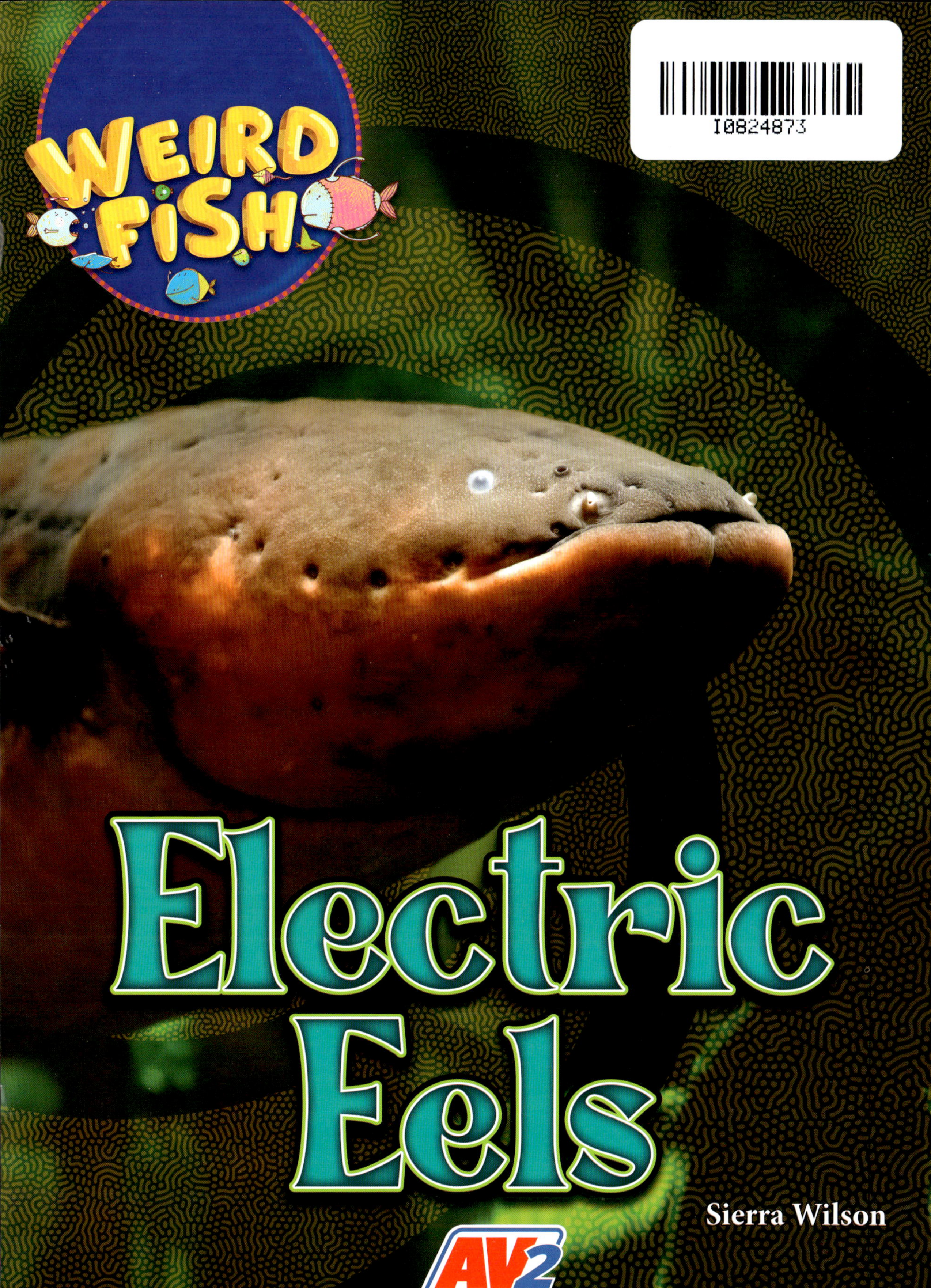

Electric Eels

Sierra Wilson

AV2
www.av2books.com

Step 1
Go to **www.av2books.com**

Step 2
Enter this unique code
JAQCFH63J

Step 3
Explore your interactive eBook!

AV2 is optimized for use on any device

Your interactive eBook comes with...

Contents
Browse a live contents page to easily navigate through resources

Audio
Listen to sections of the book read aloud

Videos
Watch informative video clips

Weblinks
Gain additional information for research

Slideshows
View images and captions

Try This!
Complete activities and hands-on experiments

Key Words
Study vocabulary, and complete a matching word activity

Quizzes
Test your knowledge

Share
Share titles within your Learning Management System (LMS) or Library Circulation System

Citation
Create bibliographical references following the Chicago Manual of Style

This title is part of our AV2 digital subscription

1-Year K–5 Subscription
ISBN 978-1-7911-3320-7

Access hundreds of AV2 titles with our digital subscription.
Sign up for a FREE trial at **www.av2books.com/trial**

Electric Eels

CONTENTS

The Electric Eel

Have you ever seen an electric fish? It may have been an electric eel. Despite their name, these South American fish are not actually eels. They are part of the knifefish group of fish. Relatives of knifefish include catfish and carp. Electric eels can produce powerful electric shocks to hunt **prey**, scare off **predators**, and find their way through their murky **habitats**.

There are three types of electric eels. They are the electric eel, Vari's electric eel, and Volta's electric eel.

Electric Eel

Scientific Name *Electrophorus electricus*

Diet Carnivore

Length 8.2 feet (2.5 meters)

Habitat Freshwater rivers, lakes, ponds

Electric Eel Features

An electric eel's body has many different **adaptations**. Some of these help the fish find food or stay safe. Others help it to survive in its habitat.

Swim Bladder

A pocket-like **organ** called a swim bladder is inside an electric eel's body. It helps the fish control its **buoyancy**.

Body
Electric eels wrap their bodies around large prey. This helps them deliver a stronger electric shock.

Mouth
Even though electric eels have **gills**, they get most of their oxygen by breathing air. Much like a person's **lungs**, their mouths have blood vessels that take in oxygen.

Skin
Electric eels have thick, slimy skin that protects them from their own electric shocks and other dangers.

Electric Eel Life Cycle

During the dry season in South America, female electric eels lay up to 17,000 eggs. Male electric eels keep the eggs safe in hidden nests made of **saliva**. The males guard their young until they are big enough to protect themselves. When the rainy season comes, young electric eels spread into new rivers, lakes, and ponds. Electric eels can live up to 22 years.

An electric eel can weigh more than 48 pounds (22 kilograms).

Knifefish Length Comparison
Adult Male Human
5.75 feet (1.8 m)
Black Ghost Knifefish
1.5 feet (0.5 m)
Aba Aba Knifefish
5.5 feet (1.7 m)
Electric Eel
8.2 feet (2.5 m)

Electric Eel Habitats

Electric eels live in the Amazon River and connected bodies of water. These fish prefer slow-moving water. They are often found in bodies of water that have muddy bottoms.

Many electric eels live in forests that flood every year. These places are called várzeas.

Electric Eel Facts

In nature, electric eels cannot be found anywhere other than South America.

On the Move

Electric eels use one very long **fin** along the underside of their body to help them swim. They wiggle this fin to move backward and forward. Electric eels move slowly. In fact, they often hold still and send out electric signals. Electric eels can sense disturbances in their **electric currents**, which helps them know when prey, predators, or possible **mates** are near.

An electric eel's tail makes up about 80 percent of its length.

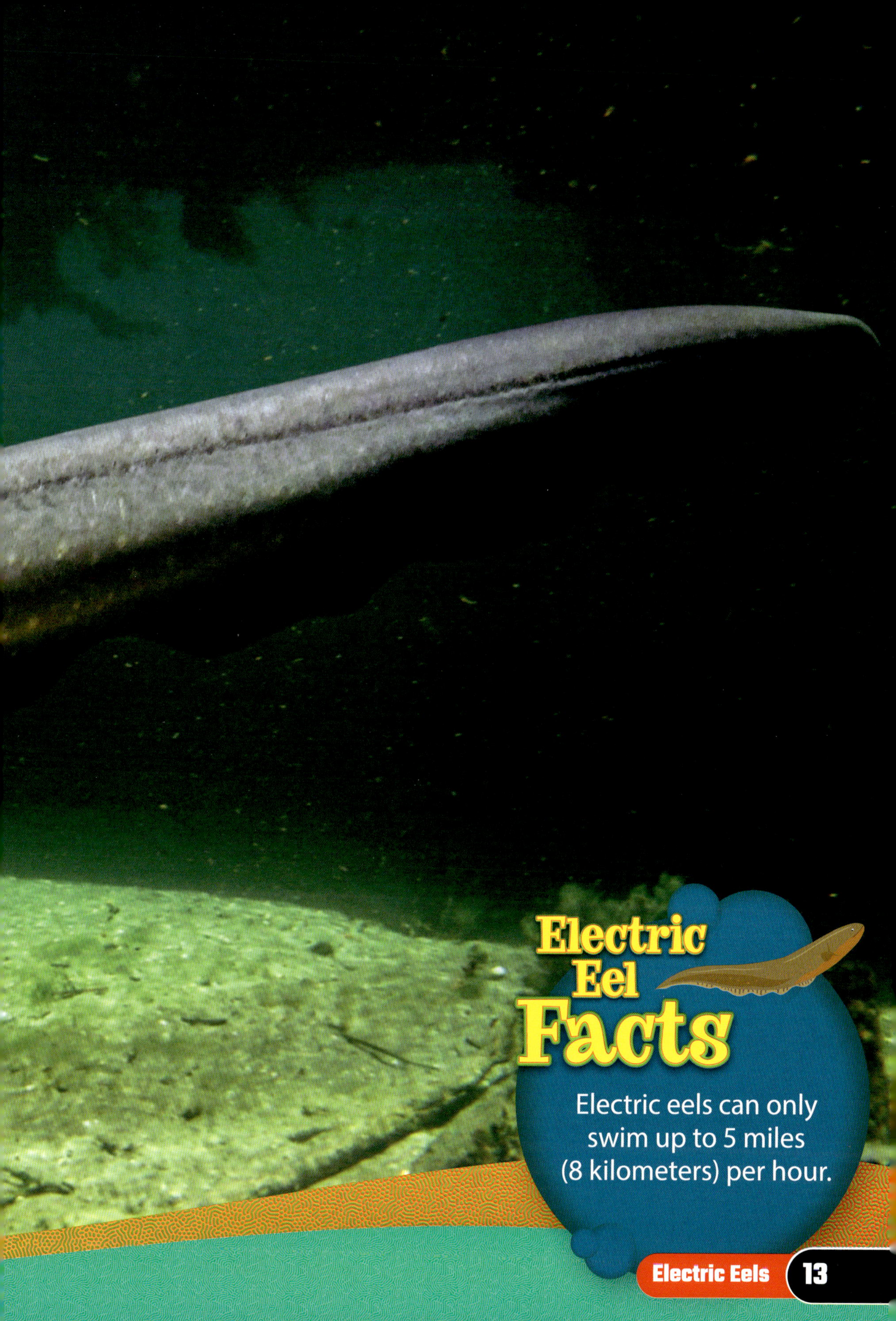
Electric
Eel
Facts
Electric eels can only
swim up to 5 miles
(8 kilometers) per hour.

What They Eat

Electric eels hunt by stunning their prey with electricity. When they sense prey nearby, they send out strong electric shocks. Then, they suck down their food whole. Electric eels mainly feed on other fish. They also eat amphibians, insects, and small mammals.

Shocking prey stops it from moving. This helps protect an electric eel's mouth from any spiny food it may eat.

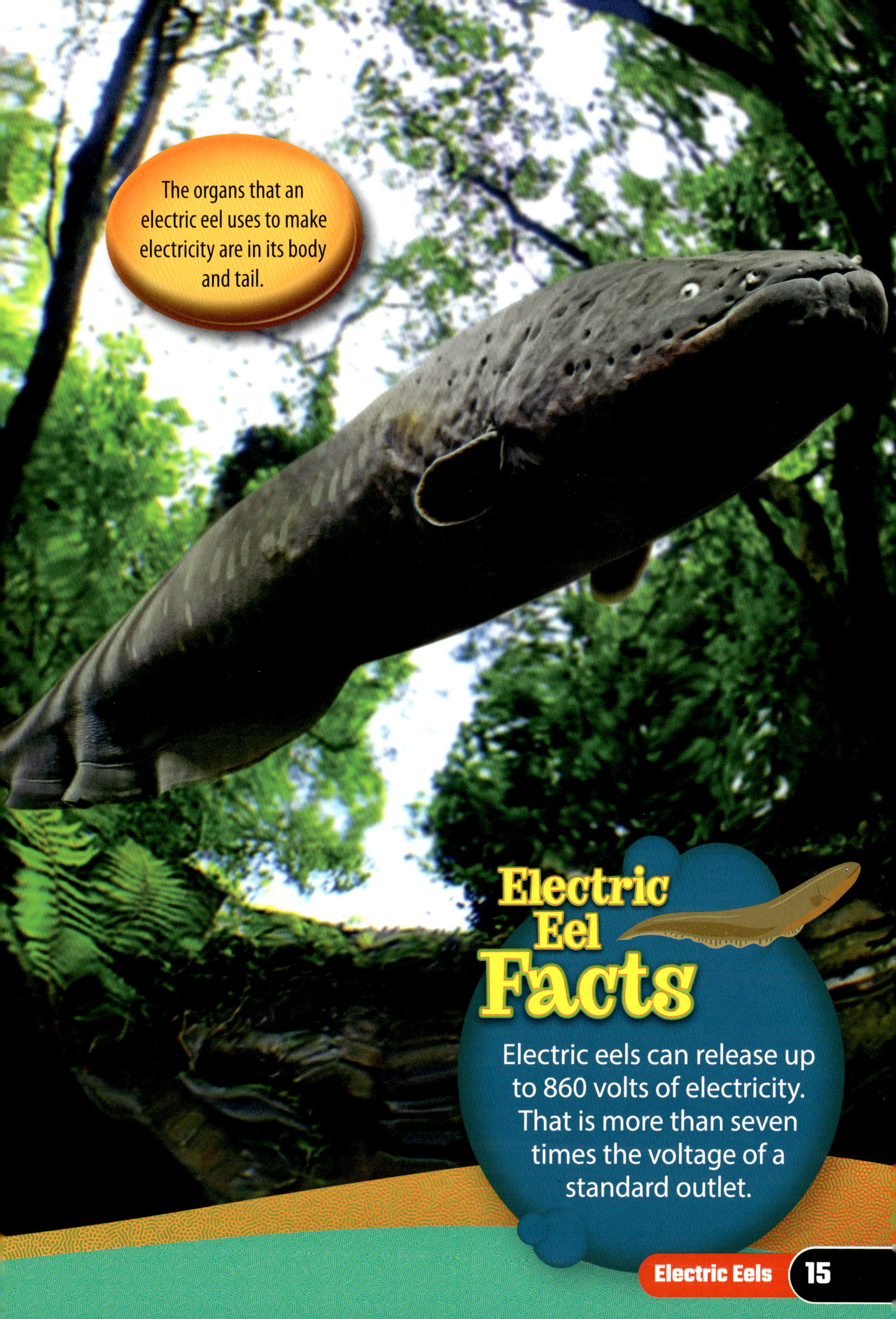

Electric Eel Facts

Electric eels can release up to 860 volts of electricity. That is more than seven times the voltage of a standard outlet.

Staying Safe

As large fish, electric eels are usually safe from predators. However, during the dry season, when water is shallower, they sometimes have to fight off large land animals. One way they do this is by lunging toward a predator to shock it directly instead of through the water. This makes the shock even more powerful.

When possible, electric eels avoid conflict. They can easily hide among plants or in murky waters.

Viewing Electric Eels

Electric eels can be difficult to see in their natural habitat because they are **nocturnal** and live in murky water. These fish are also dangerous to humans and should not be approached. They can cause pain and serious injury. People can safely see and learn more about electric eels by visiting **aquariums** and zoos.

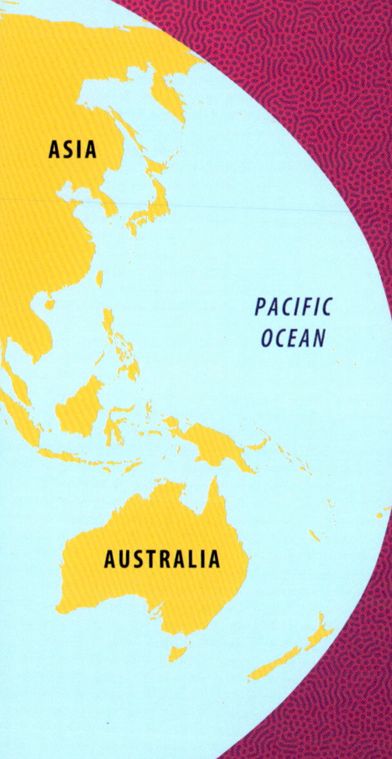

1 Amazon Rainforest, Brazil

Scientists study electric eels in their natural habitat in the Amazon region. In 2012, Dr. Douglas Bastos found more than 100 electric eels in a single small lake and discovered that they hunt in packs.

2 Chattanooga, Tennessee

The Tennessee Aquarium has an electric eel exhibit. It is connected to lights and a speaker that show the electricity level in the water. In 2019, the aquarium used this exhibit to light a Christmas tree.

3 Stuttgart, Germany

Germany's Wilhelma Zoo was originally designed to be a private residence for a king. Now, it is home to one of the most diverse collections of animals in the world, including the electric eel.

Protecting Electric Eels

Even though electric eels are not currently **endangered**, they are threatened by human behavior. As people cut down the Amazon Rainforest, they destroy the flooded forest areas electric eels call home. **Pollutants** in rivers also harm electric eels and the fish they eat. People have taken steps to keep electric eels from becoming endangered. It is **illegal** to catch one without special permission.

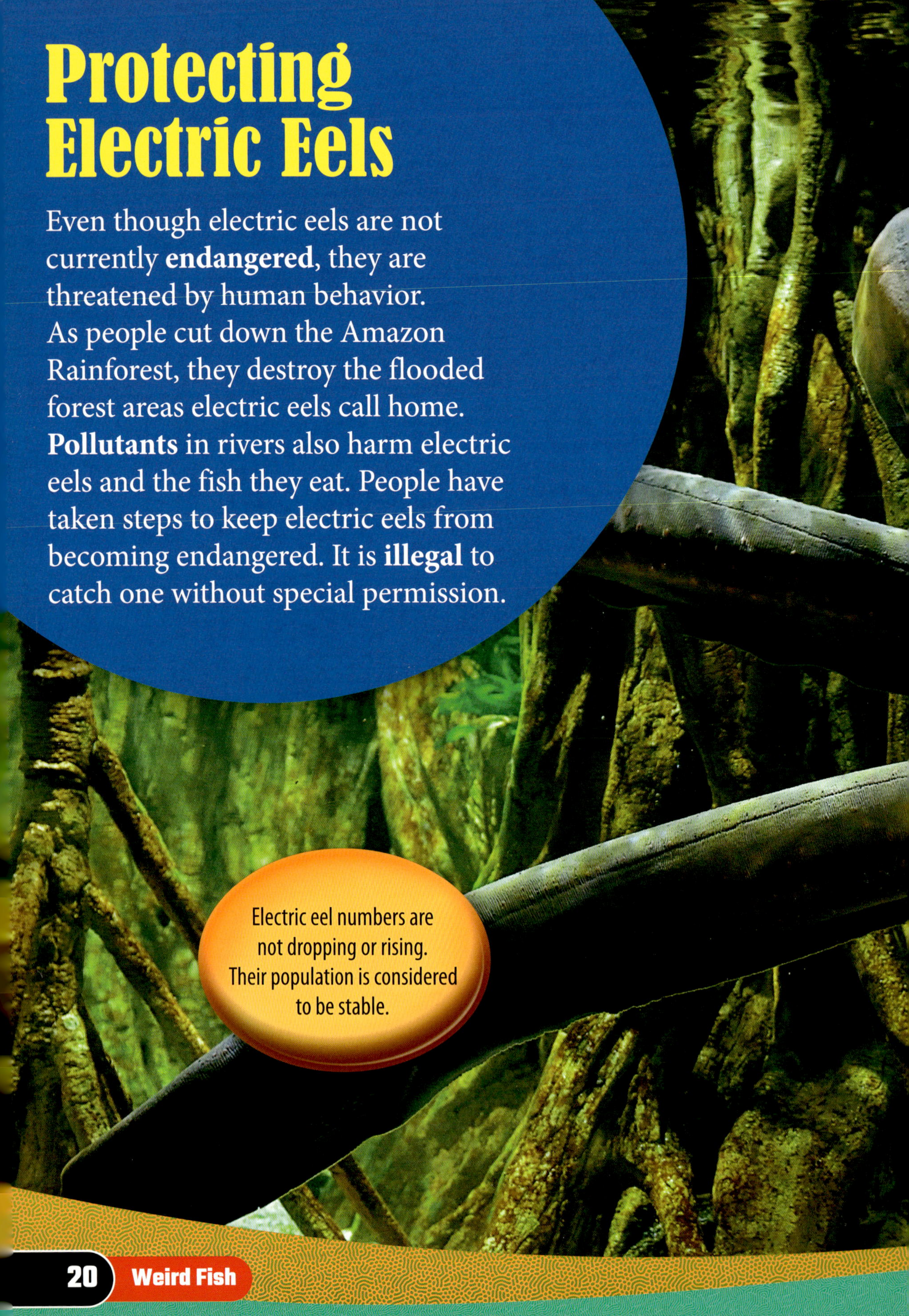

Electric eel numbers are not dropping or rising. Their population is considered to be stable.

Electric Eel Conservation Status
LC
Least Concern
NT
Near Threatened
VU
Vulnerable
EN
Endangered
CR
Critically Endangered
EW
Extinct in the Wild
EX
Extinct

Test Your Knowledge

ANSWERS
1. Knifefish 2. Three 3. In its body and tail 4. In saliva nests 5. A long fin on the underside of their body 6. By stunning prey with electricity 7. The Tennessee Aquarium 8. Special permission

Key Words

adaptations: changes in living things that make them better able to survive in their homes

aquariums: places that keep and display animals that live in water

buoyancy: the ability to float in air or water

electric currents: flowing electricity

endangered: close to disappearing from Earth

fin: a part of an animal's body that is used to help it move and steer in the water

gills: organs used by fish to take in oxygen from water

habitats: the places where plants or animals live

illegal: against the law

lungs: organs used by some animals to take in oxygen from air

mates: members of a pair of animals that can reproduce or have children

nocturnal: active at night

organ: a part of the body that performs specific tasks

pollutants: things that make air or water unclean or unsafe

predators: animals that hunt other animals

prey: animals that are hunted by other animals

saliva: watery liquid in the mouth

Index

Get the best of both worlds.

AV2 bridges the gap between print and digital.

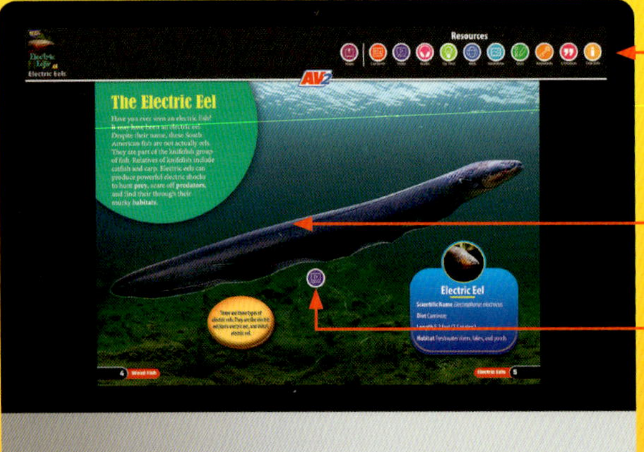

The expandable resources toolbar enables quick access to content including **videos**, **audio**, **activities**, **weblinks**, **slideshows**, **quizzes**, and **key words**.

Animated videos make static images come alive.

Resource icons on each page help readers to further **explore key concepts**.

Published by AV2
276 5th Avenue
Suite 704 #917
New York, NY 10001
Website: www.av2books.com

Copyright ©2022 AV2
All rights reserved. No part of this publication may be reproduced, stored in a retrieval system, or transmitted in any form or by any means, electronic, mechanical, photocopying, recording, or otherwise, without the prior written permission of the publisher.

Library of Congress Cataloging-in-Publication Data

Names: Wilson, Sierra, author.
Title: Electric eels / Sierra Wilson.
Description: New York, NY : AV2, [2022] | Series: Weird fish | Includes index. | Audience: Grades 2-3
Identifiers: LCCN 2021022193 (print) | LCCN 2021022194 (ebook) | ISBN 9781791142346 (library binding) | ISBN 9781791142353 (paperback) | ISBN 9781791142360 (ebook other)
Subjects: LCSH: Electric eel--Juvenile literature.
Classification: LCC QL638.E34 W55 2022 (print) | LCC QL638.E34 (ebook) | DDC 597/.43--dc23
LC record available at https://lccn.loc.gov/2021022193
LC ebook record available at https://lccn.loc.gov/2021022194

Printed in Guangzhou, China
1 2 3 4 5 6 7 8 9 0 25 24 23 22 21

062021
101120

Art Director: Terry Paulhus Project Coordinator: John Willis

Every reasonable effort has been made to trace ownership and to obtain permission to reprint copyright material. The publisher would be pleased to have any errors or omissions brought to its attention so that they may be corrected in subsequent printings.

The publisher acknowledges Alamy, Getty Images, Minden Pictures, Newscom, and Shutterstock as the primary image suppliers for this title.

View new titles and product videos at www.av2books.com